Extreme Weather

BLIZZARDS

Martha London

DiscoverRoo
An Imprint of Pop!
popbooksonline.com

abdobooks.com

Published by Pop!, a division of ABDO, PO Box 398166, Minneapolis, Minnesota 55439.

Printed in the United States of America, North Mankato, Minnesota.

052019
092019

THIS BOOK CONTAINS RECYCLED MATERIALS

Cover Photo: Shutterstock Images
Interior Photos: Shutterstock Images, 1, 5, 6, 9, 19, 22, 27, 28; NOAA, 7; Tom Stromme/The Bismarck Tribune/AP Images, 8, 31; iStockphoto, 10–11, 13, 18, 20, 23, 25, 29 (snowflakes), 29 (checkmarks), 29 (house), 29 (flashlight), 29 (bread), 29 (car), 29 (scarf), 29 (television), 29 (bottom), 30; The Herald Bulletin/AP Images, 14; Stephan Savoia/AP Images, 15; Pavel Golovkin/AP Images, 16; Sun xin/Imaginechina/AP Images, 17; Yin Bogu/Xinhua News Agency/Newscom, 26

Editor: Connor Stratton
Series Designer: Sophie Geister-Jones

Library of Congress Control Number: 2018965055

Publisher's Cataloging-in-Publication Data

Names: London, Martha, author.

Title: Blizzards / by Martha London.

Description: Minneapolis, Minnesota : Pop!, 2020 | Series: Extreme weather | Includes online resources and index.

Identifiers: ISBN 9781532163913 (lib. bdg.) | ISBN 9781532165351 (ebook)

Subjects: LCSH: Blizzards--Juvenile literature. | Winter storms--Juvenile literature. | Severe storms--Juvenile literature. | Natural disasters--Juvenile literature.

Classification: DDC 551.555--dc23

WELCOME TO

DiscoverRoo!

Pop open this book and you'll find QR codes loaded with information, so you can learn even more!

Scan this code* and others like it while you read, or visit the website below to make this book pop!

popbooksonline.com/blizzards

*Scanning QR codes requires a web-enabled smart device with a QR code reader app and a camera.

TABLE OF CONTENTS

CHAPTER 1

HOW BLIZZARDS FORM

Blizzards are strong snowstorms. But not all snowstorms are blizzards. Winds must travel 35 miles per hour or faster. People must be able to see no more than four football fields away. These high

WATCH A VIDEO HERE!

Some blizzards can last for days.

winds and low **visibility** must last at least three hours.

A blizzard needs four things to form. First, the air needs to be cold. Second, warm air needs to meet that cold air. The warm air will rise over the cold air. Third, the air must have

When warm and cold air meet, clouds can form.

enough moisture. These conditions form clouds to make snow. Finally, winds must be strong enough to blow the snow a lot.

EXTREME BLIZZARD!

In 2018, a blizzard hit the East Coast of the United States and Canada. Some places received 2 feet of snow. States as far south as Florida received snow. In New Hampshire, winds traveled more than 120 miles per hour.

Sometimes, blizzards form without any snowfall. Winds can blow snow that was already on the ground. These snowstorms are called ground blizzards.

DID YOU KNOW? **Blizzards have been recorded in every month except July, August, and September.**

In some cases, blowing snow from a ground blizzard can freeze people's eyes shut.

HOW SNOW FORMS

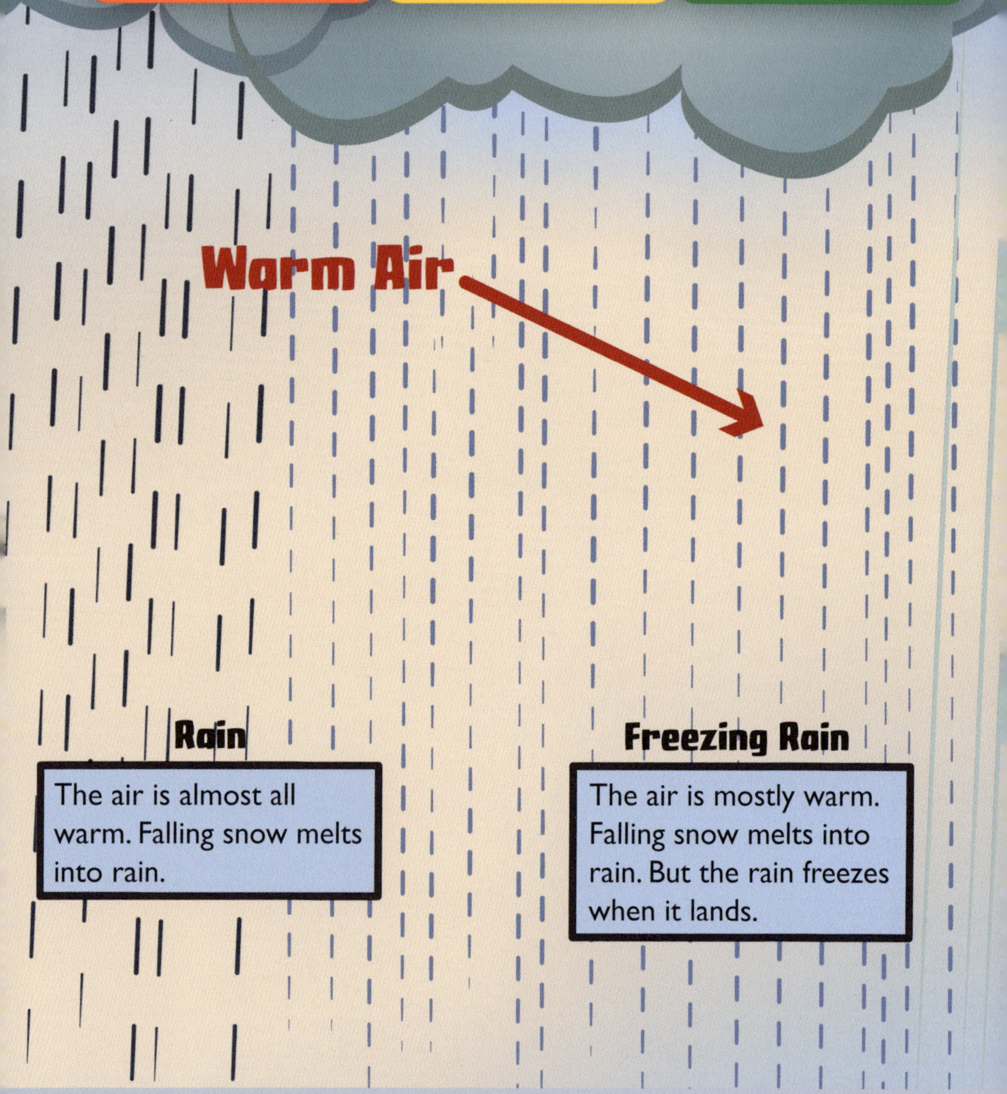

Blizzards form when cold and warm air meet. Other kinds of weather form this way too.

Sleet

The air is partly warm and partly cold. Falling snow melts into rain. But the rain refreezes into sleet on the way down.

Snow

The air is mostly cold. Snow falls to the ground. If enough wind blows, a blizzard can form.

CHAPTER 2

WHERE BLIZZARDS FORM

Many blizzards form in the northern **Great Plains** of the United States. This area's **geography** makes blizzards more likely. The land is flat. The land also does not have many trees. Cold wind flows down easily from Canada.

LEARN MORE HERE!

Ten US states and parts of Canada make up the Great Plains.

Blizzards also form near the East Coast. Cold wind travels south to the East Coast. That air meets warmer

DID YOU KNOW?

One record-setting blizzard was the Cleveland Superbomb. In 1978, it affected an area the size of Texas and Montana combined!

During the Cleveland Superbomb, blizzard winds reached 100 miles per hour.

air from the Atlantic Ocean. This mix can make strong storms. Some are **cyclones** known as nor'easters. They often cause blizzards.

Nor'easters can bring strong waves as well as blizzards.

DID YOU KNOW?

In 2018, a two-day blizzard brought nearly 2 feet of snow to Moscow, Russia.

A boy in China uses an umbrella and face mask to get through a blizzard.

Blizzards occur in other parts of the world too. Asia is often hit with winter storms. Russia and China receive severe blizzards.

CHAPTER 3

DAMAGE

A blizzard's blowing wind and snow can cause many problems. People cannot see well during a blizzard. Car accidents are more likely. People can also get lost outside, even if they are near home.

In the United States, winter storms cause nearly half a million car crashes every year.

LEARN MORE HERE!

If people are not dressed warmly during a blizzard, frostbite can damage their skin.

Blizzards can bring extreme cold. Plus, their winds make the air feel even colder. This lower temperature is called the windchill. In a blizzard, the windchill can drop below –60 degrees Fahrenheit! A windchill that low can lead to **hypothermia**. It can cause **frostbite** too. Going outside can be dangerous.

DID YOU KNOW?

Blizzard windchills can cause people to get frostbite in less than ten minutes.

Blizzards can cause other damage too. During winter, ice often forms over many things, such as trees or power lines. Ice can be heavy. If a lot of ice coats a power line, the line is easier to break. A blizzard's strong winds can cause power lines to snap. When that happens, homes can lose power.

Hundreds of thousands of people can lose power from one blizzard.

CHAPTER 4

STAYING SAFE

Meteorologists track winter storm clouds to tell if a blizzard might come. These scientists usually know a few days ahead of time if a blizzard is coming. If they expect a blizzard, they send out

COMPLETE AN ACTIVITY HERE!

a blizzard warning. These warnings give people time to prepare for the storm.

Blizzard warnings often appear on televisions and phones.

Families often buy many supplies to prepare for a blizzard.

People make sure they have enough supplies for the blizzard. They gather food, batteries for flashlights, and warm clothes. These supplies help people stay safe if their power goes out or if they cannot leave their home.

During the blizzard, people stay inside if they can. They also watch for signs of **frostbite** and **hypothermia**. Numb and yellow skin can mean frostbite. Shivering and extreme tiredness can mean hypothermia.

If people notice these signs, they try to warm up.

DID YOU KNOW?

A person with hypothermia should warm the center of his or her body first.

Blizzard Safety Checklist

- Find shelter as soon as possible.
- Gather supplies in case of a power outage.
- Avoid driving.
- Dress in warm clothing.
- Follow the news for weather updates.

MAKING CONNECTIONS

TEXT-TO-SELF

Are blizzards common where you live? Why or why not?

TEXT-TO-TEXT

Have you read books about other kinds of storms? What do they have in common with blizzards? How are they different?

TEXT-TO-WORLD

Blizzards can make it hard to see outside. Why might people try to avoid driving during blizzards?

GLOSSARY

cyclone – a storm that rotates and can bring rain or snow.

frostbite – skin damage caused by freezing.

geography – the features of Earth's surface in a specific place.

Great Plains – an area of flat land in the United States and Canada that goes east from the Rocky Mountains for hundreds of miles.

hypothermia – a condition in which the temperature of a person's body is very low.

meteorologist – a scientist who studies and predicts the weather.

visibility – how far people are able to see.

INDEX

ONLINE RESOURCES

popbooksonline.com

Scan this code* and others like it while you read, or visit the website below to make this book pop!

popbooksonline.com/blizzards

*Scanning QR codes requires a web-enabled smart device with a QR code reader app and a camera.